A Merrow Monograph

FLAME RESISTANT COTTON

Merrow Monographs

General Editor: J. Gordon Cook, B.SC., PH.D., F.R.I.C.

Concise, authoritative reviews of technical topics.

Textile Technology Series

MM/TT/1 *Decrystallized Cotton*, Leon Segal.
MM/TT/2 *Textile Printing*, L. W. C. Miles.
MM/TT/3 *Cotton Modification with Oxiranes (Epoxides)*, J. B. McKelvey.
MM/TT/4 *Physiology and Hygiene of Materials and Clothing*, E. T. Renbourn.
MM/TT/5 *Permanent Setting of Wool*, A. J. Farnworth and J. Delmenico.
MM/TT/6 *Drycleaning*, A. E. Johnson.
MM/TT/7 *Twistless Yarns*, H. J. Selling.
MM/TT/8 *Elastomeric Fibres*, R. Meredith.
MM/TT/9 *Mothproofing of Wool*, J. R. McPhee.
MM/TT/10 *Durable Press Cotton Goods*, Robert J. Harper, Jr.
MM/TT/11 *Flame Resistant Cotton*, Wilson A. Reeves and George L. Drake, Jr.

FLAME RESISTANT COTTON

Wilson A. Reeves, B.S., M.S., Sc.D.

and

George L. Drake, Jr., B.S., M.S.

Southern Regional Research Laboratory,
U.S. Department of Agriculture,
New Orleans, La., U.S.A.

MERROW

Merrow Publishing Co. Ltd.,
276 Hempstead Road
Watford Herts. England

ISBN 0 900 54142 3 Clothbound
ISBN 0 900 54143 1 Paperback

Printed in Great Britain at the Pitman Press, Bath

CONTENTS

1

INTRODUCTION

Man has been concerned for centuries with the dangers involved when cellulosic materials ignite. An early description of flameproofing cellulosic materials is to be found in a British patent granted to Obadiah Wyld in 1735. This patent described the use of "alum, borax, vitriol, or copperas" to prevent the flaming of paper pulp or textiles.

Through the years there has been an increasing interest in imparting flame resistance to fabrics, and a recognition of the need to protect lives and property against flammable cellulosic materials.

The Burning Process

Burning of cellulosic textiles takes place in two steps; (1) flaming, and (2) glowing.

(1) *Flaming*

Thermal decomposition of cellulose, which always precedes flaming, results in gaseous, liquid, tarry, and solid products. As the flammable gases burn, the liquids and tars volatilize, and some of these volatile fractions burn; other fractions yield a carbonaceous material which does not burn readily.

(2) *Glowing*

The carbonaceous residue glows until essentially all organic material is consumed and only a fluffy ash remains.

The temperature of flaming cellulose is 400–450°C, whereas the afterglow temperature is about 600°C.

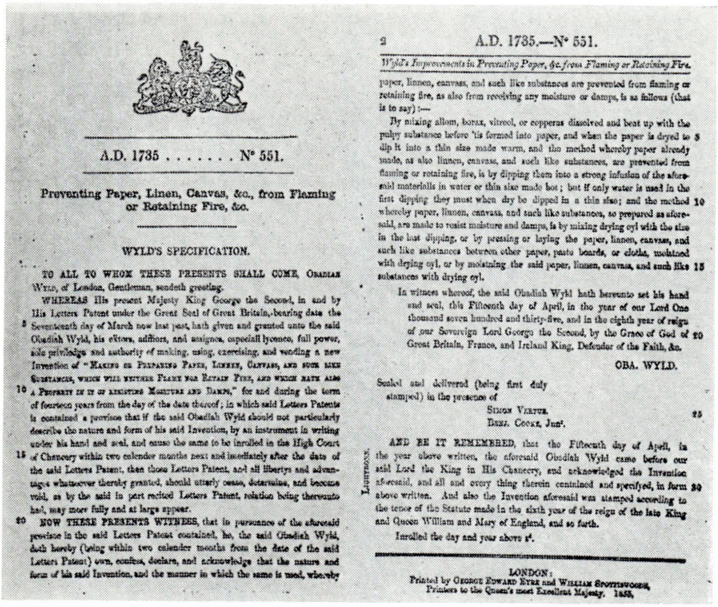

A.D. 1735 . . . . N° 551.

Preventing Paper, Linen, Canvas, &c., from Flaming or Retaining Fire, &c.

WYLD'S SPECIFICATION.

TO ALL TO WHOM THESE PRESENTS SHALL COME, OBADIAH WYLD, of London, Gentleman, sendeth greeting.

WHEREAS His present Majesty King George the Second, in and by His Letters Patent under the Great Seal of Great Britain, bearing date the Seventeenth day of March now last past, hath given and granted unto the said Obadiah Wyld, his extors, adiions, and assignes, especiall lycence, full power, sole priviledge and authority of making, using, exercising, and vending a new Invention of "MAKING OR PREPARING PAPER, LINNEN, CANVASS, AND SUCH LIKE SUBSTANCES, WHICH WILL NEITHER FLAME NOR RETAIN FIRE, AND WHICH HATH ALSO A PROPERTY IN IT OF RESISTING MOISTURE AND DAMPS," for and during the term of fourteen years from the day of the date thereof; in which said Letters Patents is contained a provisoe that if the said Obadiah Wyld should not particularly describe the nature and form of his said Invention, by an instrument in writing under his hand and seal, and cause the same to be inrolled in the High Court of Chancery within two calender months next and imediately after the date of the said Letters Patent, then those Letters Patent, and all libertys and advantages whatsoever thereby granted, should utterly cease, determine, and become void, as by the said in part recited Letters Patent, relation being thereunto had, may more fully and at large appear.

NOW THESE PRESENTS WITNESS, that in pursuance of the aforesaid provisoe in the said Letters Patent contained, he, the said Obadiah Wyld, doth hereby (being within two calender months from the date of the said Letters Patent) own, confess, declare, and acknowledge that the nature and form of his said Invention, and the manner in which the same is used, whereby

2 A.D. 1735.—N° 551.

Wyld's Improvements in Preventing Paper, &c. from Flaming or Retaining Fire.

paper, linnen, canvass, and such like substances are prevented from flaming or retaining fire, as also from receiving any moisture or damps, is as follows (that is to say):—

By mixing allom, borax, vitreol, or copperas dissolved and beat up with the pulpy substance before 'tis formed into paper, and when the paper is dryed to dip it into a thin size made warm, and the method whereby paper already made, as also linnen, canvass, and such like substances, are prevented from flaming or retaining fire, is by dipping them into a strong infusion of the aforesaid materialls in water or thin size made hot; but if only water is used in the first dipping they must when dry be dipped in a thin size; and the method whereby paper, linnen, canvass, and such like substances, so prepared as aforesaid, are made to resist moisture and damps, is by mixing drying oyl with the size in the last dipping, or by pressing or laying the paper, linnen, canvass, and such like substances between other paper, paste boards, or cloths, moistned with drying oyl, or by moistning the said paper, linnen, canvass, and such like substances with drying oyl.

In witness whereof, the said Obadiah Wyld hath hereunto set his hand and seal, this Fifteenth day of April, in the year of our Lord One thousand seven hundred and thirty-five, and in the eighth year of reign of our Sovereign Lord George the Second, by the Grace of God of Great Britain, France, and Ireland King, Defender of the Faith, &c.

OBA. WYLD.

Sealed and delivered (being first duly stamped) in the presence of

SIMON VERTUE.
BENJ. COOKE, Junr.

AND BE IT REMEMBERED, that the Fifteenth day of April, in the year above written, the aforesaid Obadiah Wyld came before our said Lord the King in His Chancery, and acknowledged the Invention aforesaid, and all and every thing therein contained and specifyed, in form above written. And also the Invention aforesaid was stamped according to the tenor of the Statute made in the sixth year of the reign of the late King and Queen William and Mary of England, and so forth.

Inrolled the day and year above wr.

LIGHTBOXES.

LONDON:
Printed by GEORGE EDWARD EYRE and WILLIAM SPOTTISWOODE,
Printers to the Queen's most Excellent Majesty. 1855.

Patent granted to Obadiah Wyld in 1735, relating to the flameproofing of cellulosic materials

Terminology

A *flame resistant textile* will not continue to support a flame once the source of ignition has been removed. Likewise, a *glow resistant textile* will not continue to glow when the source of ignition is removed. Often, a fabric is flame resistant but not glow resistant.

When both flame and glow resistance are exhibited, the product is said to be *fire resistant.* Fire resistant cellulosic textiles char but do not melt.

Burning of Cotton

The burning characteristics of cotton and other cellulosic textiles depend to a considerable extent on the chemical and thermal properties of B-glucose, the availability of oxygen, and the nature of non-cellulosic materials in the cotton. Fibre properties, such as the ignition temperature, the rate and heat of combustion, the combustion products and moisture content also affect the burning of cotton. When purified cotton burns, the quantity of char produced is very small. The use of thin-layer chromatography has enabled 19 carbonyl compounds to be identified in the tar and volatile fractions produced by pyrolysis of cotton at 420°C.[1] The most abundant were glyoxal, glycolaldehyde, pyruvaldehyde, and 5-(hydroxymethyl) furfural.

Flame Retardants

Several theories have been proposed to explain how flame retardants function; they include coating, gas, thermal, hydrogen bonding, and catalytic dehydration of cellulose to carbon and water.

Coating Theory

According to the coating theory, the flame retardant forms a glassy coating and a stable foam on the fibre surface. This protects the fabric from the air, serving as a barrier to the flame and entrapping the volatile materials evolved during combustion.[2] This theory may be adequate in respect of some of the early flame retardants, but it does not explain the action of more recent types.

Gas Theory

The gas theory suggests that the flame retardant decomposes at burning temperatures, yielding gases which do not burn; they dilute the flammable gases produced by the decomposition of the cellulose to a concentration below that at which they can burn. This theory may apply satisfactorily to the flame retardants containing halogens, but it is more likely that halogens function, in fact, through a free radical mechanism.

Thermal Theories

Two thermal mechanisms have been proposed to explain the retardation of the burning of cellulose. One states that the caloric input from a source is dissipated by an endothermic change, such as fusion or sublimation of the flame retardant, which prevents propagation of the flame. The other mechanism states that heat is conducted away so rapidly that the cellulose does not reach combustion temperatures. These two mechanisms are not applicable to most known flame retardants.

Hydrogen Bonding Theory

In the hydrogen bonding theory, it is suggested that the strong hydrogen bonding of the flame retardant helps stabilize the bonds between cellulose fragments by reducing their volatility and consequently their combustibility. This theory seems weak, however, as it is improbable that these hydrogen bonds could exist at 400°–500°C.

Dehydration Theory

The dehydration mechanism proposed by Leatherman(3) states that the path of decomposition is altered, the amount of flammable gases and tars thus being reduced to the extent that combustion will not continue once the source of the flame is removed. The most recent and widely accepted modification of this theory states that flame retardancy is brought about by catalytic dehydration of cellulose through the reaction of a Lewis acid with the cellulose via a carbonium ion mechanism.(4) Thus, reduced flammability is due to an increase in char and a decrease in the combustibles. This theory seems to apply satisfactorily to many flame retardants based on compounds and polymers containing phosphorus.

The pyrolytic products from flame resistant cotton are similar to those from purified cellulose.(5) Thus, it is apparent that reduced flammability is due to a reduction in the amount of combustibles formed, as well as to the formation of some additional decomposition products. The tar from cotton made flame resistant with a phosphorus-containing compound or polymer contains little or no laevoglucosan, whereas that from purified cotton contains laevoglucosan as a major component. With this type flame retardant, a correlation also exists between the flame resistance of fabric and the ratio of char to tar or the ratio of CO to CO_2 of thermal degradation. The higher the CO to CO_2 ratio, or the char to tar ratio, the better the flame resistance.

Glowing is retarded or prevented by chemicals that inhibit the oxidation of carbon. Almost four times as much heat is evolved when carbon is oxidized to carbon dioxide as when it is oxidized to carbon monoxide. The combustion (glowing) process cannot be self-sustaining, because of a lack of energy, when materials are present that prevent oxidation of carbon to carbon dioxide. Most sources of phosphorus are good antiglow agents.

The elements found most frequently in durable flame or fire retardants for cellulose are phosphorus, nitrogen, bromine, chlorine, and antimony. Of these elements only phosphorus and bromine are efficient when used alone. Nitrogen, chlorine, and antimony have seldom been suggested as flame retardants in the absence of other known flame retardant elements. The efficiency of phosphorus is greatly enhanced by the presence of certain nitrogen-, bromine-, and chlorine-containing compounds.

2

EVALUATION OF FIRE RESISTANCE

An ideal test should be (a) rapid, (b) simple, (c) reproducible, (d) graduated, and (e) related to end use. No test yet devised meets these requirements in full, and many organizations are presently seeking to improve the existing tests or develop new ones.

It is generally agreed that (a) ease of ignition, (b) glow time, (c) rate of flame spread, and (d) amount of heat liberated are the essential factors which determine the fire hazard of a fabric. A cellulosic fabric which is allowed to hang without restriction in a vertical position and ignited at the base by flame impinging on both sides is considered to be in the most hazardous position with respect to the testing of fire resistance. The heat lost from the surface is minimal and the flame and volatile gases from the burning fabric pass up and around the unburned fabric, encouraging further ignition. In the case of fabrics which tend to melt and drip when exposed to flames, however, portions of the fabric will melt and drop away from the suspended fabric, preventing further flaming.

Vertical Flame Test

This test has achieved the widest acceptance for determining the fire resistance (flame and glow resistance) of fabrics. It was designed specifically to measure the fire resistance of cellulosic textiles in which a flame is allowed to impinge on the bottom edge of the textile held in a vertical position.

In this test, three fabric specimens, $2\frac{3}{4} \times 10$ in., are conditioned by exposure to an atmosphere of 70 ± 2°F and

Fig. 1. Vertical flame test. The test chamber with test specimen in holder

65 ± 2 per cent relative humidity for at least 8 hr before testing. The conditioned fabric specimen is then mounted in a suitable clamp and placed in a standard cabinet, and the bottom edge of the fabric is exposed to a standard flame for exactly 12 sec under controlled conditions. After removal of the flame, time of flaming and afterglow are observed. Figure 1 illustrates the test chamber with test specimen in holder.

The length of the char produced by the combined flaming and afterglow is measured from the lower edge of the sample to the uppermost point of the charred area. The height of the charred area is not easily determined on some samples. To overcome this problem a given weight (dependent on the weight of the fabric used) is attached to the bottom corner of the burned sample by means of a hook, and the opposite corner is then lifted slowly until the weight is raised from the surface of the table. This produces a tear through the charred area to a point where the fabric can withstand further tearing. The char length can then be measured easily.

AATCC Test Method 34–1969

This is a similar vertical flame test designed for measuring the fire resistance of textile fabrics when a flame is applied to one edge. It differs in that it has been modified to distinguish between fabrics with borderline flame resistance; it is also applicable to the synthetic fabrics which melt or shrink away when exposed to a flame.

In the revised test method, specimens of the thermoplastic fibres are stitched with three rows of glass thread stitches in the long direction, the first row of stitches being directly in the centre of the fabric from one end to the other. The second and third rows of stitching are sewed on each side parallel to the first row, the stitches being $\frac{3}{8}$ in. apart. The edge to be ignited is trimmed straight across and free from loose ends. Matheson B gas is specified.

A 3-sec exposure to the flame as well as a 12-sec exposure is incorporated into the test in order to screen out the marginally-treated fabrics which would pass the 12-sec exposure but would fail the 3-sec exposure to the flame source.[6,7]

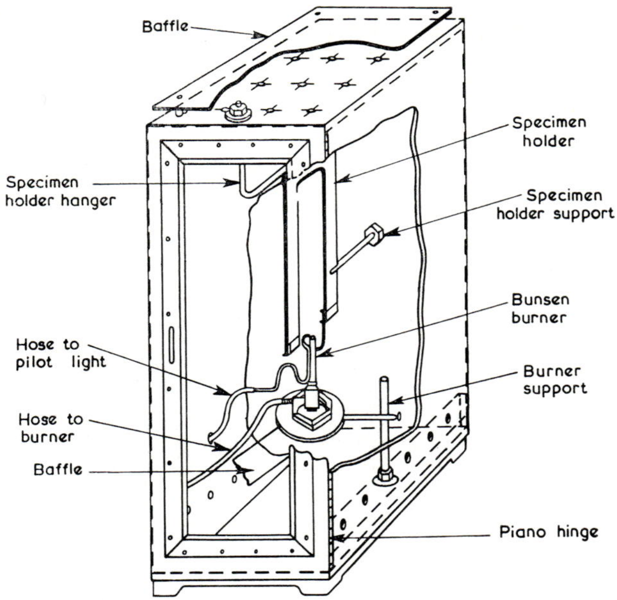

Fig. 2. Method 5903. Schematic of apparatus

Method 5903 (December 1968) Fed. Test Method STD. No. 191

This is also a vertical test method. In addition to defining the vertical position of the samples and flame exposure conditions, it defines gas composition, burner, cabinet, temperature, and humidity test conditions.

Figure 2 shows a schematic of the apparatus for Method 5903.

The fabric specimen, $2\frac{3}{4} \times 12$ in., is placed in a specimen holder, which is placed in the cabinet with the face and back of the fabric specimen facing the sides of the cabinet so that the flame can be seen on both sides of the fabric during the combustion period. The sample is exposed to the test flame within 20 sec after removal from the standard atmosphere.

Ignition in this test is for 12 sec. The time that the burning of the specimen continues after removal of the burner is recorded, as well as glow time after the flame is extinguished. To determine the char length, a certain tearing weight is applied depending on the weight of the cloth prior to treatment.[8]

ASTM Test Method D626-55T and NFPA Test Method 701

These are also vertical test methods.[9,10] NFPA Test Method 701 differs in that it consists of a large-scale as well as a small-scale test. It requires also that specimens be conditioned for 1 to 1½ hr at 140°–145°F and tested in the dry condition. This makes the test more severe. The Underwriters test[11] is very similar to the NFPA Test Method.

Vertical Suspension Flammability Tester with Flame Spread Meter

This test was developed in Switzerland utilizing a design by E. P. Martin of the Ciba Laboratories.[12] It was developed because the approved 45° angle test and clothing fires did not correlate. The test employs a vertical suspension of fabric, because most clothing fires are started more easily with fabrics draped in this position. It uses a 5 × 25 cm sample of newly finished fabric and fabric after repeated washing or dry cleaning. The upper 10 mm of a 30 mm high gas flame at standard pressure is applied to the bottom edge of a sample vertically suspended in a draught-free cabinet at room temperature for a period of 1 to 10 sec.

Using this type of tester, ignitability, combustion characteristics, duration of combustion and flame spread rate can be determined.

Nine cotton threads are strung horizontally at regular intervals over the surface of the sample, one end of each thread being attached to a recording pen which marks a continuous running tape. The moment any of the threads are burned the breakage is automatically recorded.

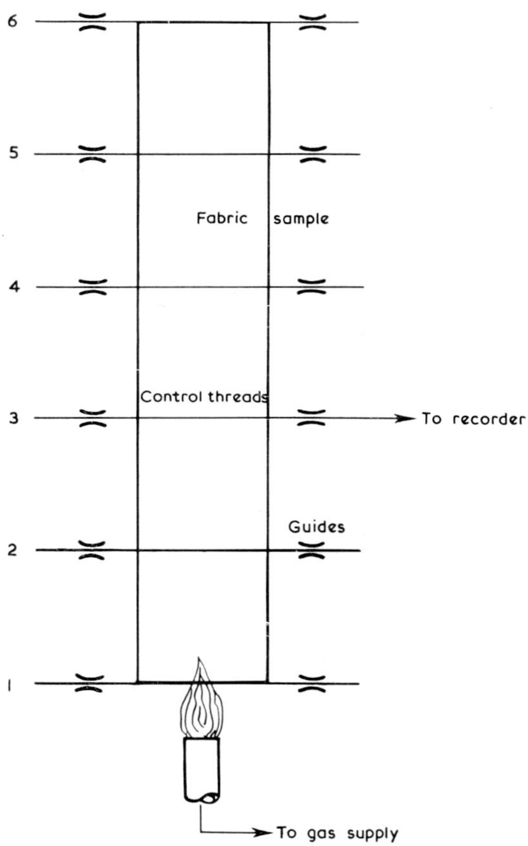

Fig. 3. Vertical suspension flammability tester with flame spread meter

The sample is placed in position, the threads and pens are connected, the tape is set in motion, and the flame is applied. Duration of the combustion process and flame spread are registered automatically by the nine control threads.

The data on the tape are evaluated by plotting them on a graph. The duration of combustion is marked in seconds on

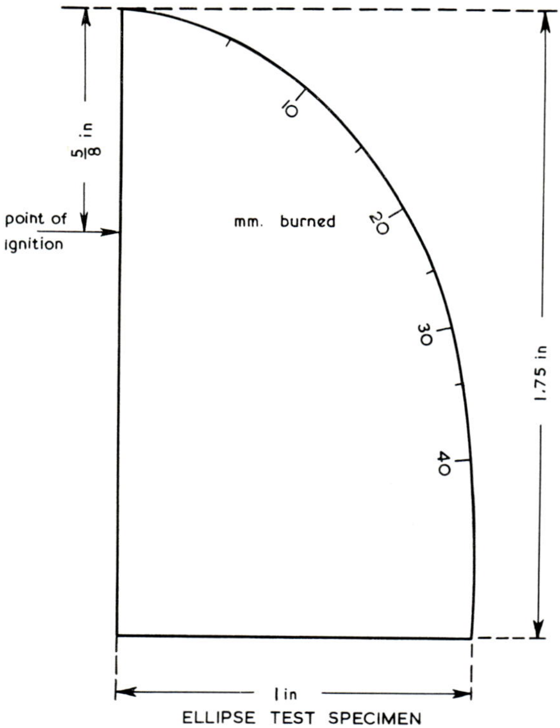

Fig. 4. Ellipse flammability test of J. P. Stevens and Co. Inc.

the x axis and char length in cm on the y axis. A diagram of this apparatus is shown in Fig. 3.

Ellipse Flammability Test of J. P. Stevens and Co., Inc.

This test was designed so that small amounts of fabric could be used to determine the degree of flame resistance of a fabric. It is somewhat similar to the vertical test, but has the advantages of requiring only a single specimen and giving a continuous gradation of results.

In this test[13] a piece of fabric in the shape of a quarter ellipse, major axis 3·5 in. and minor axis 2 in., is clamped in the vertical position and ignited at the upper edge ($\frac{5}{8}$ in. below the top) with a pin point flame. This is shown in Fig. 4. The shape of the specimen is such that if the fabric has excellent flame resistance the curved edges will not burn at all. Samples without flame resistance will ignite at the top and continue to burn down the curved edge all the way to the clamp.

With borderline fabrics, the flame will extinguish itself part way down the curved edge. The angle which the edge makes with the vertical at the point of self-extinguishment indicates the flammability; i.e., 0° = complete flammability; 90° = complete flame resistance.

Federal Aviation Agency (FAA) Flame Retardance Test

This test[14] is also somewhat similar to the vertical test. A treated specimen, $2\frac{1}{2} \times 12$ in., is clamped in a horizontal frame. One and one-half inches is allowed to drop vertically into a Bunsen burner flame (3 in. high with a $1\frac{1}{2}$ in. blue cone) for a period of 15 sec. The flame is then removed and the burning rate is measured. If the flame goes out during the first minute and the sample does not burn more than 4 in., it is regarded as being self-extinguishing. This test method is shown in Fig. 5.

G. M. Match Test Method 30–27

In this test[15] the specimen is ignited vertically. The test has come into widespread use for evaluating upholstery fabrics used in automobile interiors. Specimens 1×12 in. are preconditioned and ignited with a wooden match for 15 sec. A satisfactory flame retardant fabric is one which extinguishes itself within 5 sec after the match flame is removed. To pass the test, the sample must not burn past the 6 in. mark

Two other tests, Ford Motor Test BN 24–1 and Chrysler Motor Horizontal Test LP–463KC–13–01, are also used to evaluate the flammability of automotive interior textile trim materials.

Fig. 5. Federal Aviation Agency (FAA) flame retardance test

The Match Test

Although this is not an official test method, it is one of the simplest of the screening tests and is widely used throughout the industry.

A 1 × 7 cm strip of fabric is held by one end with the

fabric surface vertical and the end is ignited. The angular position is determined at which the fabric must be held for the flame to go out. For example, if a strip is held vertically by the bottom and the flame goes out when it is ignited at the top, it is said to have passed the 0° flame test, which requires very little flame resistance; on the other hand, if the strip is rotated through 180° and the flame goes out when the strip is held with the long edge horizontal and the fabric surface

Fig. 6. The match test

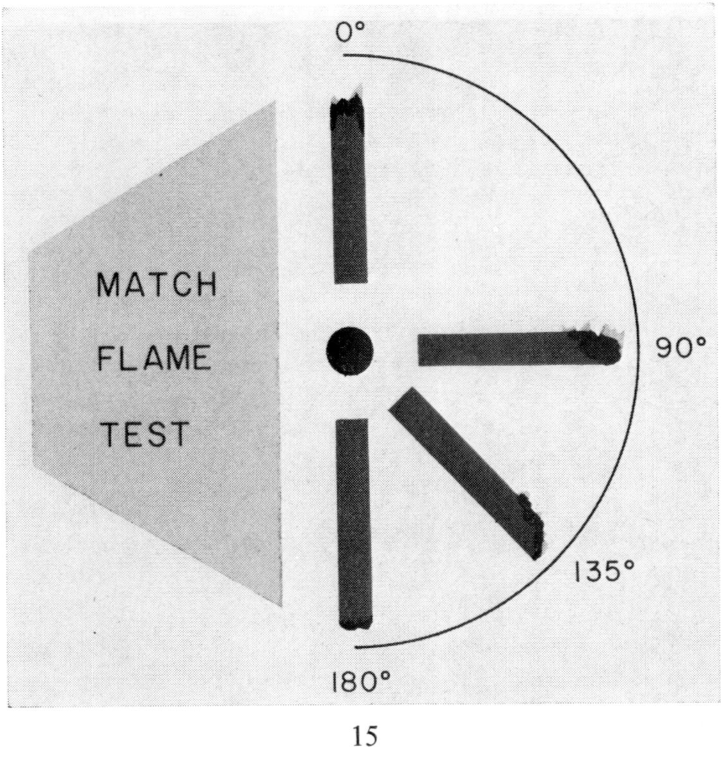

vertical, it has passed the 90° match test and is regarded as being moderately flame resistant. Strips of untreated cloth burn continuously, even when held vertically and ignited at the top (see Fig. 6).

The Match Test can provide useful qualitative information about the burning characteristics of a fabric. It registers the degree of flame resistance in terms of angular degrees, the rapidity of movement of the flame, the smoke characteristics, the intensity of the flame, glow time, and char characteristics.[16,17]

With a little practice, this test can be performed easily and with a high degree of reproducibility. It has been found that most flame resistant fabrics which pass the 135° angle will generally pass the standard vertical flame test.

British Standard 3119—Method of Test for Flameproof Materials

This method is also based on a vertical strip test. It requires a test specimen 12½ in. long by 2 in. wide which is suspended from a clip in a draught-free box with the bottom end supported on pins. A coal gas flame is applied to the end of the fabric for 12 sec and then withdrawn. The flaming time, the char area, and the amount of afterglow are recorded. The test specifies that no test specimen shall continue to flame for more than 8 sec, the afterglow shall not spread beyond the charred area, and the average char length or melt area shall not exceed 3·5 in. with no individual specimen exceeding 4·5 in. These requirements must be met before and after repeated launderings.

British Standard 2963—Methods of Test for the Flammability of Fabrics

This is also a vertical strip test. In Method A of the test, a sample 6 ft long and 1·5 in. wide is allowed to hang freely from the clip in a frame. Horizontal wires are spaced to mark off 50 in. from 10 in. up the test sample. A coal gas flame is applied to the end of the fabric for 12 sec and then withdrawn. If the sample does not ignite, or if the flame goes out before passing the lower marker wire, the sample is reported as not propagating a flame. If the flame passes the lower mark but not the upper mark, the sample is considered self-extinguishing. If the fabric continues to burn to the upper marker wire, the time in seconds it takes to burn from the lower to the upper mark is recorded and doubled to obtain the flame resistant rating. A rating of 150 sec or more is required of any sample before and after repeated laundering if the fabric is to pass the test.

Revised Draft British Standard Methods of Test for the Flammability of Textile Fabrics

These new revised draft British Standards[33] are being studied and if found satisfactory may replace existing British Standards 2963 and 3119. The revised draft consists of a three-method composite flammability test.

The apparatus consists of a rectangular box (400 × 30 × 1200 mm) designed to be relatively draught-free. Samples are suspended vertically from a clip located either 450 or 1080 mm from the floor of the box. The bottom edges of the sample are held with U-shaped clips attached to opposite sides of the box 220 mm from its floor. Butane or propane burners (design described in each method) are used to ignite the samples. Before testing, samples are conditioned 4 hr at 20 ± 2°C and 65 ± 2 per cent R. H. The atmosphere of the testing chamber should be 16–24°C and 45–67 per cent R. H.

Method A

Method A is intended for broad application to testing of textile fabrics (except floor coverings). Parameters reported include rate of flame propagation, duration and spread of afterglow and afterflame and the char length. In testing, the 900 × 75 mm sample is ignited at the bottom edge with a 40 mm butane or propane flame from a burner designed to allow no premixing of air with the gas supply. The burner is placed so that the 10 mm-long luminous flame is 20 mm below the middle of the bottom edge of the sample and is left in that position for 10 sec. Time required for the propagating flame to travel from a marker 450 mm from the bottom of the box to a marker 950 mm from the bottom of the box is recorded, and rate of propagation of flame is calculated.

Method B

Method B is intended as a confirmatory test for fabrics which pass Method A, but are suspected of having done so because they melt, drip, or curl away from the flame. This method would be applied primarily to fabrics made from thermoplastic fibres. The 300 × 350 mm sample is suspended as in Method A. A butane or propane flame (20 mm in length) from a burner designed to premix air with gas is directed horizontally at a point 50 mm above the middle of the bottom edge of the sample, with the end of the burner 5 mm from the fabric surface. The flame is applied for 5 sec and the maximum distance of horizontal or vertical damage, the time necessary for flame to travel to the sample's edge and to a reference line 250 mm upward, and the total flaming time are recorded.

Method C

Method C is intended solely for determining whether pile or simulated fur fabrics (except floor covering) promote rapid spread of flame when in transient contact with an ignition source. Samples (200 × 300 mm) are shaken to raise the pile, edges are folded back and stapled to minimize edge effects and then suspended vertically so that the lay of the pile is predominantly downward. The burner (as described in Method A) is adjusted to have a 20 mm flame, tilted horizontally and passed across the sample at a constant rate of about 300 mm/sec. Number of traverses made, number of traverses in which ignition occurred, whether surface flashing occurred, and whether surface flashing or burning damaged or destroyed the fabric are reported.

Oxygen Index Test Method

Considerable attention has been given to the interpretation of flammability in terms of the Oxygen Index (OI). This test was originally designed by Fenimore and Martin[18] at the research laboratories of the General Electric Company for use with self-supporting plastic specimens. The reproducibility of test results is reported to be very good. The apparatus consists of two sections, a gas metering system and the test column. In the gas metering system the volumetric flow of gas in each line is accurately determined before the streams are mixed and fed into the base of the test column. In the test column the gas mixture flows upward past the burning specimen, which is ignited at the top to burn in a candle-like manner. By adjusting the flow in each line to vary the oxygen concentration in the mixture, it is possible to determine accurately the oxygen index, the minimum oxygen concentration which will just support combustion of the material. A diagram of the apparatus is shown in Fig. 7.

This test has gained widespread use throughout the plastics industry. It is now an ASTM standard (D2863). More detailed information concerning the usefulness of this test and results obtained was reported by Isaacs.[19]

Tesoro and Meiser[20] extended the use of this test method to textiles. They evaluated the flammability of textile material as a function of chemical composition and showed that the OI is primarily governed by chemical composition, with only a small effect attributed to the weight and construction of the

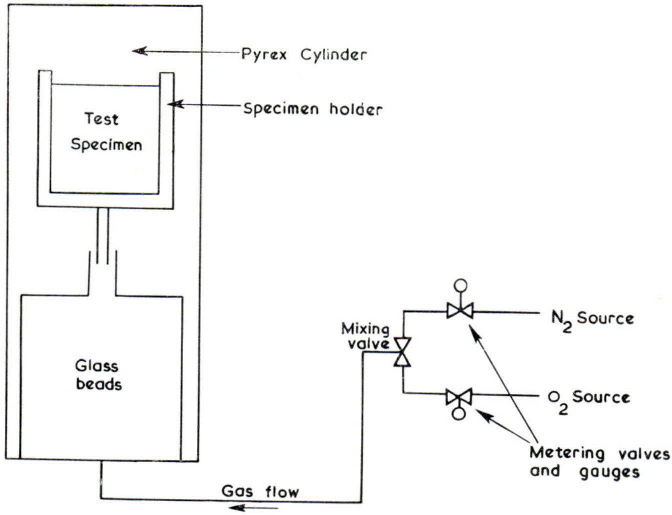

Fig. 7. Schematic of oxygen index apparatus

fabric. As shown by Tesoro *et al.*, fabrics with reduced flammability had higher OI values. Some of the values obtained by Tesoro are shown in Table 1.

Willard and Wondra[21] reported on the quantitative evaluation of the flammability characteristics of cotton and flame retardant cotton using the OI technique. According to these authors, the synergistic interaction of *N* and *P* can be described and measured. By this technique they found that reduced flammability was dependent on increasing values of *N* or *P*, and no single definite *N*/*P* ratio produced the best synergistic effects. Some relationship of the OI value to the standard vertical flame test was reported. Generally, an OI value of 0·26 was shown to be equivalent to a 5-in. char length when measured by the standard vertical flame test.

Some emphasis is now being placed on obtaining flame retardants for textiles which do not produce smoke or toxic gases during combustion.

The Michigan Chemical Company designed a smoke densitometer which can be used in combination with the GE Flammability Index Tester. They also installed an additional pyrometer in the chamber of the smoke densitometer to measure the temperature of the gases rising from the combustion chimney while the sample was being burned in the OI apparatus.[22]

TABLE 1

Limiting Oxygen Indices

Acrylic	0·182
Acetate	0·186
Polypropylene	0·186
Rayon	0·197
Cotton	0·201
Nylon	0·201
Polyester	0·206
Wool	0·252
Flame Retardant Cotton	0·270
Nomex	0·282

Circular Feed Flame Propagation Method

This is not yet an official test, but Miller[23] has shown it to have features lacking in other test methods. By using this test, flame propagation, flame temperature, and relative ease of extinction of the textiles being evaluated can be determined. In this test, fabric is mounted on the perimeter of a pair of connected wheels in such a way that the speed of rotation can be adjusted. The fabric is ignited, and the speeds of rotation of the wheels are adjusted to keep the flame from the burning fabric stationary at the highest point of the circular path. During testing, the wheel apparatus is mounted inside a large 7-in. diameter chimney. The OI values are obtained. A diagram of this apparatus is shown in Fig. 8.

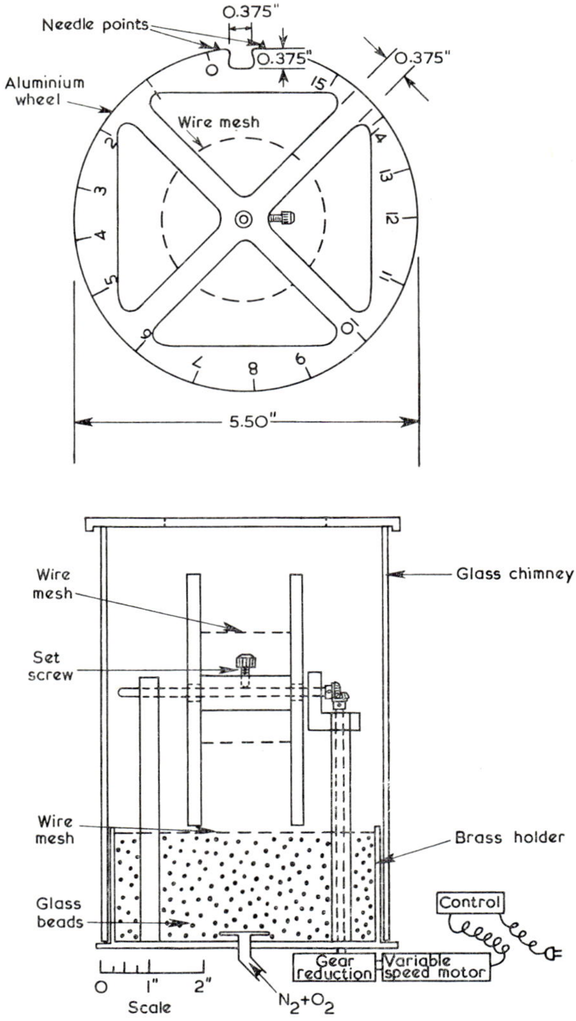

Fig. 8. Circular Feed Flame Propagation Apparatus.

Top: front view. Bottom: side view

Semicircular Test Method

Interest has been shown by Tesoro[24] in the semicircular test method. This type of test was the basis of an earlier British standard for use in evaluating flame resistance of fabric.[25] Schon[26] subsequently carried out work on this test, which was further developed by the New England Section of the AATCC for an intersectional contest paper.[27] In the test, a semicircular frame 7 in. in diameter is used for holding a 10-in. strip of fabric, and the strip is ignited at one end. The rate of burning can be measured.

Flammability Test Method

AATCC Test Method 33,[28] devised to assist in determining flame resistance of untreated fabric, was used in the draft of commercial standard CS 191–53,[29] which was the basis of the Flammable Fabrics Act.[30] The AATCC method was revised in 1962 to be in line with CS 191–53.

In this test method 2 × 6 in. test specimens are evaluated. The specimens are dried then brushed if the fabric contains a raised or napped surface. Each is placed in a specimen rack then in a special ventilated draught-free cabinet housing a standard ignition mechanism and automatic timing device. The prepared sample is held at an angle of 45° and a standard flame applied to the surface near the lower end for a period of 1 sec. The time required for the flame to travel a distance of 5 in. is recorded. Ignition or fusing of base fabrics having a nap surface is also noted as an indication of the intensity of burning. Other similar test methods are Federal Specification CCC–191b Methods 5908 and 5910.[31] (See Fig. 9.)

Rugs and Carpet Standards

On December 3, 1969, there was published in the Federal Register (33FR1792) a notice of findings that a flammability standard, which included labelling, may be needed for carpets and rugs. On December 17, 1969, the U.S. Department of Commerce issued the first proposed standard under

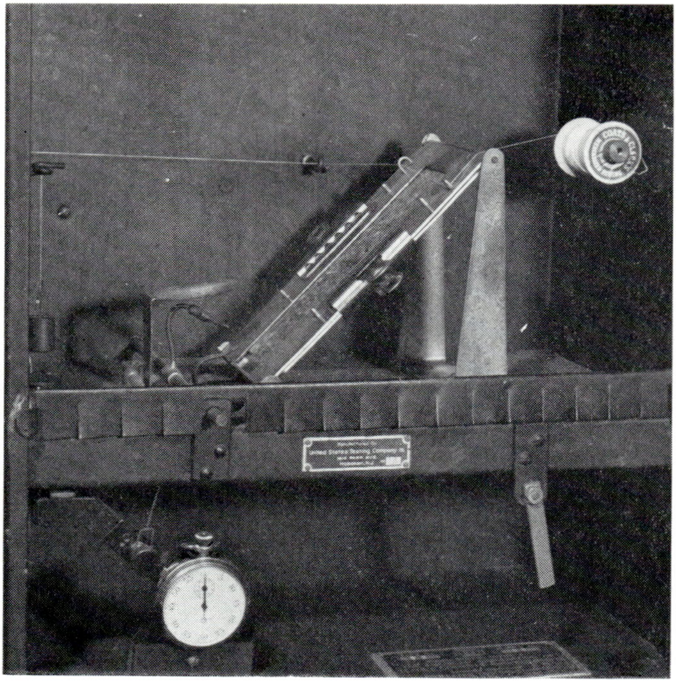

Fig. 9. Federal Specification CCC-191b Methods 5908 and 5910

the amendment to the Flammable Fabrics Act.[32] The test method is a modification of Federal Specification DDD–95, Carpets and Rugs, Wool, Nylon, Acrylic, and Modacrylic. The method tests the resistance to flammability of various rugs and carpets under draught-free conditions by requiring a maximum specified dimension of charred area when subjected to a controlled standard ignition source. The standard ignition source is a No. 1588 methenamine tablet supplied by Eli Lilly and Company, Indianapolis, Indiana. The method involves the exposure of 8–9 conditioned specimens of a rug or carpet to an ignition source in a draught-free atmosphere and measurement of the area of char. Prior to exposure to the ignition source, a steel plate with an 8-in. hole cut out of the centre is placed on the rug specimen. A rug passes the test if the charred areas of 8 specimens do not extend to within 1 in. of the edge of the hole in the steel plate.

3

NON-DURABLE FLAME RETARDANTS

Non-durable flame retardants are suitable for use on cellulosic fibres which are not to be used outdoors. The chemicals wash out of textiles when the goods are exposed to the weather, leached, or laundered. Articles should be treated again after each laundering. Even goods which are not laundered should be retreated about once each six months, as many of the non-durable-type retardants crystallize on the surface of a fabric after having diffused from the inside of the fibre. Non-durable flame retardants should be thought of as temporary retardants. They are generally unsatisfactory on polyamides, polyesters, and acetates. Like most other retardants they are more effective on cotton than on rayon.

Water-soluble non-durable flame retardants have some advantages over other flame retardants in that frequently they can be purchased from shops or local chemical supply houses in small quantities for home use. In addition, the cost of the chemicals is low and they can be applied to textiles in several ways, e.g. by dipping, spraying, and sprinkling. Since most non-durable retardants are water-soluble, they are applied from a water solution. However, some are soluble in organic solvents and under certain circumstances it might be desirable to apply these from solvents.

Material to be treated should be dry. If the material does not wet easily, about 0·1 per cent of a wetting agent should be added to the flame retardant formulation. Regardless of the technique used to apply the flame retardant, the objective is to obtain an add-on which will be sufficient to render the material glow and flame resistant. To do this most efficiently it is important that the retardant be uniformly distributed throughout a fabric. Drying of treated fabrics must be done with care. Many of the retardants will hydrolyse or decompose at temperatures above about 140°C. The decomposition may result in degradation of cellulosic textiles and/or loss of flame retardancy. In order to avoid these problems, drying should be done below about 130°C.

Many non-durable flame retardants have been considered over the years but, for various reasons, only a few have been used commercially. Ammonium salts such as ammonium chloride and ammonium sulphate, for example, are apt to cause tendering of treated fabrics during storage. The deliquescent nature of many flame retardants has kept them from being used. This deliquescent property generally results in the migration of the flame retardant from the interior to the surface of the fibre, where it may be removed easily by gentle mechanical action. Some salts are good flame retardants, but textiles treated with them have excessive afterglow. Sometimes a combination of two retardants is better than a single retardant. For example, zinc chloride is a good retardant, but it is very deliquescent. This fault is essentially eliminated when it is used in combination with zinc phosphate dissolved in an excess of ammonium hydroxide.

Examples of Non-durable Retardants

The following non-durable flame retardants may be regarded as examples of effective materials.

1. *Borax-Boric Acid*

This retardant formulation is made by preparing a paste of seven parts of boric acid in a minimum amount of water, then adding this to a water solution of three parts of borax.[34] The solution should be essentially clear, and it should be warmed if it becomes cloudy or jelly-like from standing. The amount of this retardant needed on a given textile must be determined by actual application to small samples. About 10 per cent add-on by weight is adequate for a medium-weight cotton fabric. Fabrics treated with this retardant are flame resistant but have afterglow, which can persist from about 30 sec up to several minutes.

2. *Borax-Diammonium Phosphate*

This flame retardant formulation is made by dissolving equal parts of borax and diammonium phosphate in water. It is somewhat less effective on a weight basis than borax-boric acid. However, the presence of the phosphate provides glow resistance. Borax-diammonium phosphate can cause some corrosion of metals used in the application of the treating solution to fabrics.

3. *Diammonium Phosphate*

The fire retardancy imparted by diammonium phosphate is about equivalent to that of Example 2. Diammonium phosphate provides both flame and glow resistance. It is particularly good for fabrics already containing a nitrogenous wash-wear or durable-press treatment.

4. *Ammonium Sulphate*

Ammonium sulphate is dissolved in water to form a clear solution, then ammonium hydroxide is added to give a faint odour of ammonia. This formulation also provides flame and glow resistance. It may cause a slight reduction in the strength of treated fabric.

Many additional non-durable flame retardants are available. Some are based upon alkyl and aryl phosphates of various types. Others are based upon the products obtained by reacting urea or other amides with phosphoric acid and various phosphoric acid derivatives to produce complex water-soluble materials. These are applied to cotton and rayon fabrics from a water solution.

4

SEMI-DURABLE FLAME RETARDANTS

Semi-durable flame retardants may be defined as those which withstand one or more laundry cycles, the upper limit being about 15 mild laundry cycles. Flame retardants of this class are useful in drapes, party dresses and other items which will not be laundered many times. Their main advantage over the durable type, which will be discussed later, is cost. Many semi-durable retardants were discovered unintentionally; they have resulted from research intended to develop durable flame retardants.

A. **Cellulose Phosphates**

Considerable effort has been devoted to the development of laundry-resistant, flame retardant finishes by attaching phosphates to cellulose via esterification reactions. There are many references in the literature to both cellulose sulphate and cellulose phosphate, but the methods of producing these esters are not generally applicable to textiles.

Joseph Bancroft and Sons Company were pioneers in the production of phosphate derivatives of cotton,[(35)] the process becoming known as the Ban-flame process. In the original process, a solution of urea and phosphoric acid was padded onto fabric, and the treated fabric was heated at elevated temperatures for about 5 min. A later and more practical process was based upon the use of a solution of diammonium phosphate and urea. The fabric was padded, dried, and heated at 175°C for about 7 min. By this technique 10–15 per cent esterification is achieved easily, but cellulosic fibres are severely tendered during the process.

It was later found that the tensile loss of treated fabrics could be kept to about 15 or 20 per cent.[36] The better tensile properties were obtained by careful control of curing conditions and by using formulations which would fix about 3 per cent phosphorus in the fabric. The cure extended to 1 or 2 hr at 130°C or 5 min at 175°C.

Phosphorus attached to cellulose in this fashion is very resistant to removal by boiling 1 per cent sodium hydroxide solution; however, the fabric loses its flame resistance after such treatment. The loss of flame resistance after treatment with alkali is due to the formation of sodium salts of the phosphate. The resulting salt is thermally stable to the extent that it does not decompose at flaming temperature and is not, therefore, a flame retardant. Flame retardancy can be regained by rinsing the alkali-treated fabric in a dilute solution of hydrochloric acid or other strong acids to remove the sodium ions. Soaking the acid-washed fabric in ammonium hydroxide solutions will replace the nitrogen lost originally by the alkaline treatment.

The flame resistance imparted by the urea phosphate method is durable to water, but repeated laundering with soap or with water containing calcium or magnesium salts causes the treated fabrics gradually to become flammable. This is apparently due to the fact that the cellulose ester acts as a base exchange material and forms the metal salt. Rinsing laundered fabrics with dilute ammonium chloride solutions reforms the ammonium salt of the ester and restores the flame resistance to the fabrics.

One of the earliest modifications of the Ban-flame process was to use cyanamide instead of urea with the phosphoric acid or phosphoric acid salts.[37] This modification causes less tendering of fabric. Subsequent improvements have used other amines, such as guanadine, dicyandiamide, and methylolmelamines. Other modifications use, in place of urea, a mixture of melamine or diguanadine thiosulphate with methylol-melamine.[38]

B. Ammonium Phosphate-Dicyandiamide-Formaldehyde

A mixture of methylol dicyandiamide and ammonium phosphate will polymerize to form insoluble resin in cellulosic fibres when applied to the fabric and baked. Unfortunately, at baking temperatures of about 150°C, ammonium phosphate decomposes and the phosphoric acid liberated causes excessive tendering of cellulosic fabrics.

An ingenious modification of this process developed by Bowen and colleagues,[39] consists in applying the ammonium phosphate-methyloldicyanamide mixture to the textile, then precipitating the resin by treatment with alkali. Subsequent drying and heat curing causes polymerization of the resin without excessive tendering of cellulosic fabric. The finish is resistant to water and mild washing. Fabrics treated in this way have been used for upholstery applications.

In a somewhat related process a precondensate of urea and ammonium phosphate can be insolubilized with formaldehyde to produce semi-durable flame resistance.

C. Cyanamide-Phosphoric Acid Process

The use of cyanamide and phosphoric acid for the production of flame retardant finishes was investigated in 1968.[40] The cyanamide used was a commercially available product which has relatively good stability in aqueous solutions. This work has led to the production of flame resistant fabrics with better properties than had been available using the earlier techniques.

In the new process a solution of cyanamide and phosphoric acid is padded onto fabric, which is then dried and cured. Optimum results are obtained when the solution contains about three moles of cyanamide per mole of phosphoric acid. The total solids in the treatment bath affects the durability of treated fabrics. The greater the solids content, the greater is the resistance to laundering; with 45 per cent solids in the bath, treated cotton and rayon fabrics withstand about 7–10 laundry cycles using Tide as the detergent.

The curing temperatures can be varied widely. For example, 8 min at 65°C or 2 min at 140°C is adequate. There is an exothermic reaction during drying and curing, which occurs about 2 min after fabric is in the oven. At 140°C the temperature rose by more than 120°. Fabrics cured at 140°C and at temperatures as low as 105°C were very laundry resistant; fabrics cured at 65°C, where the exothermic effect was much smaller, readily lost their flame resistance during laundering. Fabrics cured at 105°C or higher also exhibited resiliency.

About one-half of the nitrogen is durably fixed and resistant to ion exchange. The other half of the nitrogen is cationic and base exchangeable with cations such as sodium and calcium. Cotton goods subjected to this treatment retained about 60 per cent of their original strength. This is one of the main disadvantages of the process. Rot resistance and dimensional stability are increased.

D. **Phosphoryl Amides**

Semi-durable flame retardants closely resembling those made from urea and phosphoric acid discussed above can be made by reacting ammonia with phosphorus oxychloride.[(41)] The resulting phosphoryl amides are water-soluble materials consisting perhaps of some monomer but probably mostly of polymeric material. Aqueous solutions of the phosphoryl amides are applied to fabric by a pad-dry-cure process. The fabric is cured for about 5 min at 150°C.

Breaking and tear strength losses caused by this treatment are very similar to those resulting from the urea phosphoric acid treatments. Treated fabrics withstand many launderings with de-ionized water or de-ionized water containing a detergent. When the fabrics are washed in solutions containing sodium, calcium, or magnesium salts, these ions are exchanged for the amine groups and the finish is no longer a flame retardant.

5

DURABLE FLAME RETARDANTS

There are two general types of durable flame retardants for cotton: (a) those used on apparel and household fabrics, and (b) those used on industrial fabrics. Flame retardants of the first group should be unaffected by cleaning processes, provide a supple fabric, and impart little or no adverse properties. Phosphorus-containing compounds and polymers are the most important materials in this group. Retardants for industrial fabrics should be resistant to sunlight and rain. A combination of antimony oxide (Sb_4O_6) and halogen provides the most suitable type of retardant for industrial fabrics.

A. RETARDANTS FOR APPAREL AND HOUSEHOLD TEXTILES

Phosphorus-containing compounds and polymers are generally the most effective flame retardants with respect to durability to laundering. In addition, they are unique in that they prevent glowing of the substrate.

The efficiency of these flame retardants is generally affected by other elements or groups which may be present. Of the elements studied, nitrogen and bromine appear to contribute most to phosphorus-containing flame retardants. When some compounds or polymers containing nitrogen or most materials containing bromine are included, the amount of phosphorus needed to impart a given degree of flame retardancy is reduced. Generally, 1 per cent phosphorus in the presence of about 5 per cent nitrogen imparts excellent flame and glow resistance to cotton fabrics. In the absence of nitrogen or bromine, about 2 to 4 per cent phosphorus is required. The chemical nature of the nitrogen compound has a significant influence on the effectiveness of the nitrogen.

Sources of nitrogen are not all equally effective in increasing the flame resistance contributed by phosphorus.[42] This was illustrated by applying "THPOH" to aminoethyl-, carbamoylethyl-, and cyanoethyl-cotton. Samples of the three chemically modified fabrics, with unmodified controlled fabrics, were treated with various amounts of the "THPOH" and dried. Evaluation for flame resistance was by the match test.[43] The control fabric contained phosphorus but no nitrogen.

The aminoethyl cotton had the greatest degree of flame resistance per unit of phosphorus. The cyanoethyl cotton had substantially less flame resistance than the control fabric. This same relationship was observed when the chemically modified fabrics contained only 0·4 per cent nitrogen.

When the fabrics were subjected to pyrolysis, much of the phosphorus in the treated cyanoethylated cotton fabric was volatilized, whereas nearly all of the phosphorus remained in the ash when the other fabrics were pyrolysed. Since phosphorus generally contributed flame resistance by promoting dehydration of the cellulose, it is effective only in the substrate. This explains the reduced flame resistance of the treated cyanoethylated cotton fabric.

Three overlapping methods have been used in the application of durable flame retardants to cotton fabrics for apparel and household uses: (1) coating of cellulosic fibres, (2) chemical reaction with the cellulose, and (3) polymer formation within the fibres. A few of the preferred processes are discussed in this section.

(1) **Coating of Cellulosic Fibres**

Coating involves the application of the flame retardant to the surface of the fibres. This is accomplished by applying the compounds or polymers from organic solvents or aqueous emulsions, usually by the padding process.

Best results from surface deposits are obtained by using thermoplastic polymers. The quantities of polymer coating required for flame retardancy usually cause some stiffness, particularly on light or medium-weight cotton fabrics. The technique is satisfactory on cotton fabrics weighing about 250 g or more per square metre.

Bromoform Adduct of Triallyl Phosphate

This flame retardant was prepared by reacting triallyl phosphate with bromoform in an aqueous emulsion.[44] The reaction proceeds by a free radical mechanism in the presence of potassium persulphate, producing a stable emulsion of a telomer or polymer containing about 8 per cent phosphorus and 40 per cent bromine. The telomer contains some unsaturated groups and can be further polymerized by heat after application to cotton fabric, to produce durable flame retardancy. The emulsion is applied to the cellulosic material by conventional pad-dry-cure techniques. Cotton fabric requires about 18 to 20 per cent of the retardant to pass the vertical flame test after repeated laundering.

Bromoform Adduct of Allyl-Phosphazines

These have been of interest as flame retardants partly because they contain a high percentage of phosphorus and nitrogen, both of which are desirable in flame retardants.

A polymer or telomer was made by reacting bromoform with hexaallyl phosphonitrilate.[45] First, the allyl ester was prepared by the reaction of phosphonitrilic chloride with sodium allylate. Then the telomer or bromoform adduct was prepared in apueous emulsion in the manner described above for triallyl phosphate. The emulsion was applied to cotton fabric by conventional pad-dry-cure procedures. Fabric with about 14 per cent of the phosphazine passed the vertical flame test after repeated laundering. Closely related products, allyl phosphazines, are put in viscose to produce flame resistant rayon.[46]

Other polymers for use on the surface of cellulose fibres are based upon diallyl chloromethylphosphonate and diallyl cyanoethylphosphonate.[47] These monomers were polymerized with benzoyl peroxide as the catalyst. Better flame retardance was obtained, however, when the monomers were partially brominated and then polymerized. The phosphonate polymers were applied to fabric from an organic solvent or mixtures of solvents. A dry add-on of about 25 per cent of the brominated polymer was required on cotton fabric weighing about 250 g/m^2 to pass the vertical flame test. The high percentage of carbon and hydrogen per unit of phosphorus and bromine probably serves as fuel, and may account for the high add-on required for flame retardancy.

(2) **Chemical Reactions with Cellulose**

Cotton can be made flame resistant by using simple substitution reactions to form cellulose derivatives. Two types of reactions have been used for this purpose, (a) esterification and (b) etherification.

Esterification

Cellulose phosphate was the first flame resistant ester of cotton to be examined. It was made by applying a solution of phosphoric acid and urea to cotton fabric and heating at elevated temperatures.[48] A later and more practical method was based on diammonium phosphate and urea.[49] Cotton fabrics treated by these methods, in addition to suffering rather severe strength losses, underwent cation exchange during repeated laundering to form metal salts; in consequence, they lost their flame retardancy.

In 1968, an improved retardant was reported; it is based on cyanamide and phosphoric acid.[50] For best results, the cellulosic material is treated with an aqueous solution containing about three moles of cyanamide per mole of phosphoric acid, and then dried and cured. Cotton fabric treated by this process has good flame retardancy and better physical properties than cotton subjected to earlier techniques. However, the tensile strength loss is considerable and there is some ion exchange during laundering.

When phosphorus oxychloride is reacted with ammonia, the resulting phosphoryl amides are water-soluble products which contain $P(O)(NH_2)_3$ monomer and some polymeric material

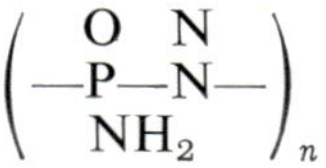

Flame retardants have been obtained by applying aqueous solutions of these materials to cotton by a pad-dry-cure process. In the reaction with cellulose, ammonia is lost during ester formation. After repeated laundering of treated fabric, the finish undergoes ion exchange with salts in the water and loses its flame retardancy.[51] Finishes with greater resistance to laundering have been made by including tetrakis (hydroxymethyl) phosphonium chloride, THPC, which can react with the phosphoramides and with cellulose.[52]

A compound closely related to phosphoryl amide imparts durable flame retardancy to cellulosic fibrous materials. This compound, chloromethyl phosphonic diamide ($ClCH_2P(O)(NH_2)_2$), is prepared by the reaction of chloromethyl phosphonic dichloride with ammonia.[53] It is applied from water solution by a pad-dry-cure method. During the reaction some ammonia and chlorine are eliminated, but the finished fabric also contains nitrogen and chlorine and much larger quantities of phosphorus. Presumably, some polymerization takes place within the fibres, but it is highly probable that esterification is extensive. Esterification results in the loss of ammonia. Fabric containing 3·2 per cent phosphorus passes the vertical flame test after three 30-min soap-soda washings at the boiling temperature. Apparently, this finish is more stable to hydrolysis than that produced by reacting phosphoryl amide with cotton.

Etherification

The 2-phosphatoethyl ether of cotton was prepared by reacting the disodium salt of 2-choroethylphosphoric acid with the cellulosic hydroxyl groups in the presence of a strong base.[54] A typical procedure for carrying out the reaction is to impregnate cotton fabric with a water solution containing about 10–15 per cent of the sodium salt and about 15–20 per cent sodium hydroxide, and then heating at 80 to 110°C for 5 to 40 min. The resulting cellulosic product is in the form of the sodium salt and is capable of cation exchange.

The cellulose derivative in the form of its metal salts does not have flame retardancy. However, when the sodium salt is converted to the free acid or ammonium salt, flame retardancy is obtained. Cotton print cloth containing 2 per cent phosphorus, when converted to the ammonium salt, will be self-extinguishing when a strip of the fabric is held in a vertical position and lighted at the bottom.

Phosphonomethylated cotton was prepared as in the manner described above, except that the disodium salt of chloromethylphosphonic acid is used.[55] This derivative has an advantage in that the carbon to phosphorus linkage is much more hydrolytically stable than the oxygen to phosphorus bond in the phosphatoethyl cellulose. For this reason, it is possible to carry out the reaction with cellulose at higher temperatures for shorter periods of time. As in the case of phosphatoethyl cellulose, it was necessary to convert the derivative to the free acid or ammonium salt to obtain flame retardancy. On cotton fabrics, these treatments generally do not reduce breaking strength, and tearing strength is reduced only slightly; however, they increase stiffness of fabric.

N-*Methylolamides of phosphines and phosphine oxides* react with cellulose fibres to crosslink cellulose molecules; they impart both flame retardancy and resiliency.[56] It is highly probable that polymers are formed in the fibre. Two compounds of this type are the methylol derivatives of tris(2-carbamoylethyl)phosphine, $P(CH_2CH_2C(O)NH_2)_3$, and tris(2-carbamoylethyl)phosphine oxide, $P(O)(CH_2CH_2C(O)NH_2)_3$. About 3·2 moles of formaldehyde per mole of the phosphorus compound was used in the methylolation. Water solutions of the methylol derivatives were applied to cotton fabric with a catalyst such as zinc nitrate or magnesium chloride. Good wrinkle recovery and flame retardancy were imparted to the fabric with about 20 per cent on weight of fabric.

Dialkylphosphonocarboxylic amides have been introduced commercially as flame retardants for cotton.[(57)] The most satisfactory finish was obtained with the monomethylol derivative of dimethylphosphonopropionamide ($(CH_3O)_2P$-$(O)CH_2CH_2C(O)NHCH_2OH$). In the presence of an acid catalyst and at about 150°C the compound reacts with cotton through the methylol group. Treatment was carried out in the presence of methylated methylolmelamine (a fixer), urea (a buffer), and an ammonium chloride catalyst. The formula was applied to cotton fabric from aqueous solution by a pad-dry-cure technique.

Good hand and flame retardancy were obtained at about 23 per cent weight gain on a fabric weighing about 230 g/m^2. Fabric strength and abrasion resistance were well within tolerable limits. This finish has achieved some success as a flame retardant for cotton fabric in the textile industry. The flame retardant is durable to repeated alkaline laundering.

Tesoro, Sello, and Willard[(58)] prepared cellulosic derivatives resembling very closely those based on the dialkyl phosphonopropionamides. Fabric was treated with three N-hydroxymethylhaloacetamides (chloro, bromo, and iodo) from a dimethylformamide (DMF) solution by a pad-dry-cure technique using zinc nitrate as the catalyst. These derivatives were then reacted in solvent with trimethyl phosphite at about 140–150°C, employing the Arbuzov reaction. The reaction rates with the phosphite decreased in the order iodo bromo chloro.

The reactions are as follows:

$$\text{CellOH} + \text{HOCH}_2\text{NHCOCH}_2\text{I} \xrightarrow{\text{DMF}} \text{CellOCH}_2\text{NHCOCH}_2\text{I} \xrightarrow{(\text{CH}_3\text{O})_3\text{P}} \text{CellOCH}_2\text{NHCOCH}_2\text{P(O)(OCH}_3)_2 + \text{CH}_3\text{I}$$

Phosphorus contents above 1·5 per cent provided good flame resistance which was durable to laundering. Loss of fabric strength was not significant.

Chance and Moreau[59,60] and Aenishanslin and co-workers[57] attached the dialkoxy phosphinyl group to the triazine ring, rather than to an alkyl group. The 2,4-diamino-6-diethoxyphosphinyl-1,3,5-triazine (DAPT) was prepared by the reaction of 2,4,6-tris-(diethoxyphosphinyl)-1,3,5-triazine with ammonia in ethanol as shown in the equation below.

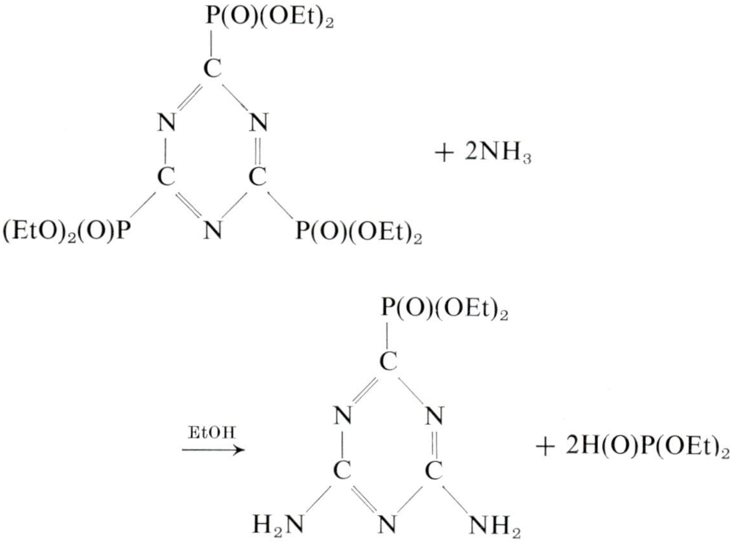

DAPT reacts with formaldehyde to form a derivative which polymerizes readily to an insoluble crosslinked polymer. Fixers such as the methylol triazines are not needed. Fabric was treated by a pad-dry-cure procedure. It was wetted with the solution to a wet pickup of about 100 per cent, dried for 5 min at 85°C, cured 5 min at 150–170°C, and washed and dried. Best results were obtained by using a solution at pH 6·6 (no catalyst).

Fabrics containing from 17·5 to 20 per cent resin add-on passed the vertical flame test even after 35 laundry cycles. Tearing strength losses were about 35 per cent and tensile strength losses about 18 per cent. Although the treated fabrics did not yellow on chlorine bleaching and scorching, some strength loss resulted, indicating chlorine retention.

(3) **Polymer Formation Within the Fibres**

This method differs from the coating method in that the chemicals are applied to the fibres as monomers or low molecular weight polymers so that they penetrate the fibres. They are subsequently polymerized to insoluble forms which are resistant to removal by laundering. In many instances, reactions also occur with the cellulose of the fibre.

This method has been the subject of considerable research during recent years. Cotton fabrics with excellent flame retardancy, hand and durability have been obtained.

Retardants Based on THPC

Tetrakis(hydroxymethyl)phosphonium chloride, THPC, is a water-soluble, crystalline compound produced in high yield by reaction of formaldehyde with phosphine and hydrochloric acid.[61,62] The methylol groups in THPC react with amides and amines to form highly insoluble polymers when the nitrogenous compounds are di- or polyfunctional. Although the phosphonium structure exists in THPC, the final product is in the form of a phosphine oxide. This phosphine oxide structure is very stable to hydrolysis. The ionic chlorine in THPC is lost during reaction and is not present in the final product. The methylol groups of THPC react with amines and amides much as N-methylol compounds react with the same amines and amides. THPC also reacts with cellulose, but the reaction is very slow under the usual conditions of producing flame resistant polymers in cotton.

THPC *Amide Process.* When these THPC-based polymers are formed inside the fibres of cellulosic fabrics in suitable concentrations the textile is rendered flame resistant.[63] The first method for using these polymers in the production of flame resistant cotton is now referred to as the THPC-amide process. A typical formulation contains 17 per cent THPC, 10 per cent methylolmelamine, 10 per cent urea, and 1 to 4 per cent of a base. The formulation is applied to fabric in a pad and cure method. This process (and slight modifications of it) have been used commercially in the U.S. for several years.

The original THPC-amide flame retardant formulation has been modified to reduce cost and improve the properties of the treated fabric. Some of the more important modifications are described below.

Antimony Oxide and Organic Chlorine Compounds

Modified formulations containing antimony oxide and organo-chlorine compounds have been used extensively.[64,65,66] The organic chlorine compounds are generally polyvinyl chloride or chlorinated paraffin. Although the antimony oxide and organic chlorine compounds are cheaper than THPC, these materials are less effective on a weight basis. When they are substituted for part of the THPC-amide formulation, greater total add-ons must be applied to the fabric to give any specified degree of flame retardance.

Allyl Phosphazine-$CHBr_3$ *Adduct*

The bromoform adduct of the allyl ester of phosphonitrilic chlorides can be produced in an aqueous emulsion using a peroxide catalyst.[45] This emulsion can be added to THPC-amide formulations. Since the adduct is a very efficient flame retardant, less of the combined formulation is required than of the THPC-amide. This modification has not been used commercially because of increased overall cost.

Allyl Phosphate-$CHBr_3$ *Adduct*

The bromoform adduct of triallyl phosphate, BAP, is also produced as an aqueous emulsion which can be added directly to THPC-amide formulations.[44] This adduct is a good flame retardant but not quite as efficient as the THPC-amide retardant. BAP and the allyl phosphazine-$CHBr_3$ adduct coat the surface of fibres instead of penetrating as THPC-amide does.

Tris (2,3-Dibromopropyl) Phosphate

The THPC-amide retardant was modified for use on polyester/cotton sheets.[67] By using an appropriate emulsifier, tris(2,3-dibromopropyl)-phosphate can be added directly to THPC-amide formulations. Unlike the chemicals used in the above three modifications, tris(2,3-dibromopropyl)-phosphate penetrates both cotton and polyester fibres to impart flame resistance.

Proban Process

In Europe the preferred commercial method of using THPC is by first reacting it with less than one mole equivalent of urea to produce a water-soluble compound that is stable in solution.[68] This compound is applied to fabric in a pad-dry process, then the fabric is exposed to ammonia and ammonium hydroxide, consecutively, to produce an insoluble flame resistant polymer inside cotton and rayon fibres. About 10 to 12 sec exposure to ammonia is adequate, and much less time is needed in the ammonium hydroxide.

Conversion of THPC *to* THPOH

Reactive methylol phosphorus compounds can be made from THPC. One way of doing this is to react 1 mole of THPC with 0·8 to 1 mole of sodium hydroxide. The product of this reaction in solution has been referred to as "THPOH".[69] This product is not yet adequately characterized but apparently consists mainly of $(HOCH_2)_3P$, THP, and CH_2O. When THPC is reacted with sodium hydroxide or other bases to produce THPOH, the pH of the solution must not exceed 7·5–7·8.

"THPOH" reacts with amines and amides to produce polymers similar to those produced by the reaction of THPC with nitrogenous compounds.

THPOH-NH_3 *Flame Retardant*

This flame retardant is applied by padding fabric with a solution of THPOH and various auxiliaries, then partially drying the fabric, and finally exposing the partially dried fabric to ammonia gas.[70] A typical formulation for flame-proofing an 8-oz cotton fabric contains 30 per cent THPOH, wetting and softening agents. The ammonia is forced directly through the fabric, and reaction is complete within a few seconds.

Since the THPOH-NH_3 flame retardant does not stiffen fabric, the process is applicable to most cotton fabrics. Fabrics weighing as little as 55 g/m^2 have been flameproofed satisfactorily by this process. Another outstanding feature is strength; it is usually greater, sometimes 25 per cent greater than the untreated fabric.

THPOH-*Amide Flame Retardant*

In applying this retardant, the reagents are padded onto the fabric, then dried and cured at about 150°C for 3 min. The typical formulation required on medium weight cotton fabric is 17 per cent THPOH, 10 per cent urea, and 10 per cent trimethylolmelamine.(71) With the THPOH-amide retardant, fabrics retain about 80 to 90 per cent of their original breaking strength. Stiffness is less than that imparted by the THPC-amide process. Because of the reduced stiffness, this retardant is applicable to a wider variety of fabrics including, for example, sheeting. Wrinkle recovery (W+F) is about 270°.

THPOH-NH_3 *Copper Complex*

Donaldson and Daigle(72) modified the THPOH-NH_3 cure process and devised a simplified pad-dry-cure process for the application of the THPOH-NH_3 flame retardant to cotton. They utilized the complexing ability of copper salts with phosphorus compounds ammonia, and amines. Copper salts, such as cupric nitrate, cupric sulphate, or cuprous chloride, when added to the THPOH solution in a 1:4 mole ratio of copper salt to THPOH, yield a system which is stable in the presence of ammonium hydroxide. This stabilization is probably due to complex formation between the Cu and phosphorus atoms. Without the Cu salt, the NH_4OH reacts immediately with the THPOH on mixing to form a white, insoluble polymeric material.

The process consists of padding the fabric with the solution and curing at about 150°C for 3 min. The finish is durable to laundering and has very little effect on the hand or strength of the fabric. The copper present imparts a slight bluish colour to the fabric. Methylol compounds, such as trimethylolmelamine, can be added to the treating solution to replace a part of the THPOH, thereby producing a lower cost durable flame retardant finish for cotton.(73)

THPC-*Cyanamide Flame Retardant*

A durable flame retardant finish for cotton, based upon THPC, cyanamide and phosphoric acid, has been developed by Normand, Donaldson, and Drake.(74) Fabric is treated by padding it through a water solution of the three components, then drying and curing at elevated temperatures. The flame retardant formulation imparts a high degree of flame resistance to fabric.

An 8-oz sateen fabric with as little as 12 per cent add-on passed the vertical flame test after 30 laundering cycles. About 15 per cent add-on was required for print cloth to have this durability. Treated fabrics have a good hand and appearance. Losses in breaking and tearing strengths associated with this treatment were about 25 and 50 per cent respectively. Chlorine bleaching alone did not discolour any of the treated fabrics. There was very slight yellowing after chlorine bleaching and scorching. The flame retardant finish did not impart any colour to the fabric, and when treated samples were exposed to a flame very little odour or smoke was produced. This last property might be of great benefit in confined areas where fumes could be a hazard during a fire.

Flame Retardants Based on APO

Tris(1-aziridinyl)phosphine oxide, APO, can be made in good yield by the reaction $OPCl_3 + \underline{CH_2CH_2NH} \rightarrow OP(\underline{NCH_2CH_2})_3$. APO is soluble in water and many organic solvents. It is toxic and must be handled with care. When APO in cotton is heated with acid or base it can react with cellulose, homopolymerize, and copolymerize with other compounds which contain active hydrogens. When used alone it imparts flame resistance and durable press properties.[75] After washing treated fabrics, they apparently do not exhibit toxic properties.

APO-THPC. A mixture containing equal weights of APO and THPC produces a good flame retardant for cotton.[76] These compounds copolymerize and react with cellulose when fabric is padded, dried, and cured at about 150°C for 4 min. Treated fabrics are supple and retain about 80 per cent breaking and 60 per cent tearing strength. The finish is exceedingly durable to repeated laundering.

APO-*Diammonium Phosphate.* Diammonium phosphate can serve as a catalyst and copolymerize with APO to produce an efficient flame retardant.[(77)] An aqueous solution containing about 5 per cent of the phosphate and 18 per cent APO is adequate to flameproof cotton fabrics weighing about 250 g per square metre. The finish is applied by a pad, dry, and cure procedure, curing at 150°C for 5 min.

APO-*Thiourea.* Fabrics treated with a solution containing 24 per cent APO and 12 per cent thiourea are rendered flame resistant when the impregnated fabric is dried and cured for about 2 min at 160°C.[(78)] The finish causes about 20 per cent loss in breaking strength and 40 per cent loss in tearing strength. It also imparts some resiliency.

Effect of Metal Ions Picked Up By Flame Retardants During Laundering

It has long been known that ion exchange properties in a flame retardant could be detrimental to an otherwise good and laundry-resistant retardant. The phosphorus in cotton resulting from the reaction of a mixture of urea and ammonium phosphate with fabric at 150°C is durable to repeated launderings, but the flame resistance decreases progressively. This is caused by a gradual loss of nitrogen from the complex cellulose derivative, with a simultaneous gradual build-up of calcium, magnesium, or other metal salt. Pickup of only 1 per cent metal ion is adequate to reduce flame resistance significantly. Flame resistance is regained when the metal ions are replaced with hydrogen or ammonium ions.

It has been suggested that the phosphate contributes flame resistance by producing acid groups which esterify cellulose and subsequently promote dehydration during pyrolysis of treated cellulose. It was reasoned that because the calcium and magnesium salts are thermally stable they are incapable of esterifying cellulose and, therefore, do not contribute flame resistance. Thus, the absence of cation exchange properties in a flame retardant has been considered to be essential if the retardant is to contribute flame resistance after repeated laundering in hard water.

Most detergents for household and commercial laundering contain phosphates. These phosphates react with calcium and magnesium ions adsorbed or contained within cellulosic fibres to produce insoluble salts. When the insoluble salts are produced in the wash water, excess phosphate from the built detergent suspends the salts, and they cause few problems. But the excess phosphate does not remove all the insoluble salts formed within the fibre. The build-up of insoluble salt is gradual, and it is influenced by the hardness of the water and the nature of the finish on the textile.

Insoluble phosphates can cause erroneous interpretations of the durability of a flame retardant when phosphorus analyses are used as the index.[79] The phosphorus content of a flame resistant fabric can increase substantially during repeated laundering. The phosphorus content of a fabric containing a phosphonate-type flame retardant increased from an initial 1·89 to 4·0 per cent during 50 laundry cycles in water at pH 10·5 with a "built" detergent.[80] Other flame resistant fabrics gained less phosphorus, but gained enough to cause serious errors if phosphorus content is used to establish the durability of the flame retardant.

A rinse in dilute acid will remove the detergent phosphorus and this should be done before testing most flame retardant fabrics. Calcium phosphate affects flame resistance of phosphorus-containing retardants nearly as much as calcium acetate. It might be predicted that the harmful effects of calcium would be offset by the contribution made by phosphate ions. The phosphorus in flame retardants generally makes its contribution to flame and glow resistance in the substrate rather than in the vapour phase. The calcium ion, which is non-volatile at flaming temperatures, must limit its ill effects to the substrate.

To check this point, calcium acetate was added to fabric containing ammonium bromide which acts primarily in the vapour phase. Flame resistance as measured by the match test was 120° and 160° with 4 and 5 per cent ammonium bromide, respectively. The addition of 1 per cent calcium acetate did not reduce resistance in either case. It appears from this experiment that calcium and similar metal ions are not detrimental to flame retardants which act mainly in the vapour phase.

Phosphorus analyses of flame resistant fabrics laundered with soaps which do not contain phosphates are generally a good measure of the durability of the flame retardant. The actual flame resistance of laundered fabrics, however, is generally decreased even though the phosphorus content remains essentially constant. These fabrics pick up calcium and magnesium soaps during laundering. When the soap is extracted with acid, flame resistance is regained. Reduction in flame resistance is due mainly to the calcium and magnesium in the soap, but also to a lesser extent to the added fuel provided by the fatty group.

B. FLAME RETARDANTS FOR INDUSTRIAL FABRICS

The flame retardant in fabrics used for tents, awnings, tarpaulins, and for other outdoor purposes must be resistant to actinic degradation and the other elements of weather. Several inorganic flame retardant compounds are resistant to actinic degradation, but they are either too expensive to apply to fabric or they are not durable to leaching. The degrading effects of weather on flame resistant fabrics varies, depending upon the season of the year, location, latitude, humidity, temperature, atmospheric contaminants, and a number of less important factors. The best and only reliable way to evaluate flame resistance of industrial fabrics is to expose samples of the fabric to the weather in the location where the fabric is to be used. The degree of flame resistance after exposure is determined by standard flame tests.

Antimony Oxide-Chlorinated Paraffins

Antimony oxide, Sb_4O_6, is essentially inert when placed in cellulosic fibres and exhibits little or no flame resistant properties.

The presence of a halogen with antimony oxide serves to activate both components and yields flame retardant systems with moderate efficiency.(81) When a cellulosic fibre containing antimony oxide and an appropriate halogen-containing compound is heated, a halogen acid is produced which reacts with antimony oxide to form antimony oxychloride. The antimony oxychloride decomposes at about 170°C to form antimony trichloride which is known to be a good flame retardant. Flame retardant processes based upon antimony oxide and organic halogen compounds have been used mostly in the production of weather resistant goods.

Several sources of halogen are satisfactory for use with Sb_4O_6, the preferred sources being chlorinated paraffins, polyvinyl-chloride, and polyvinylidenechloride. Most frequently, a combination of these sources of chlorine is used. The antimony-halogen system does not eliminate afterglow It is necessary to add a phosphate or borate to achieve this. A paraffin containing 70 per cent chlorine imparts considerable stiffness to fabric. Excessive stiffness is avoided by using two parts of paraffin containing 40 per cent chlorine with one part of paraffin containing 70 per cent chlorine.

A typical formulation for flameproofing cotton duck contains 7 per cent Sb_4O_6, 20 per cent mixed chlorinated paraffin, 4 per cent $CaCO_3$, 1 per cent aluminum stearate, 1 per cent pentachlorophenol, 9 per cent pigment, 7 per cent resin binder, and 51 per cent organic solvent. The finish is applied to cellulosic fabrics, generally cotton, by a pad-dry technique using conventional finishing equipment resulting in weight add-ons of 40–50 per cent.(82) The calcium carbonate in this finish neutralizes any hydrochloric acid that might result from actinic degradation of the chlorinated paraffins. Aluminum stearate in combination with the chlorinated paraffins provides water repellency and the pentachlorophenol provides mildew resistance.

Outdoor exposure studies of a treated cotton duck fabric indicate that very little change occurs in the fabric during a 12-month period. The reduction in breaking strength during one year's exposure is generally about 30 per cent. Much less strength is lost during subsequent years of exposure. Perhaps the greatest disadvantages of this type of retardant are the high add-on requirements and the fact that the finishes stiffen fabrics. The use of polyvinyl chloride instead of part of the chlorinated paraffins reduces the stiffness.

Antimony-Titanium Complex

Although titanium and antimony oxides when used alone contribute very little flame resistance in the absence of a halogen, a mixture of titanyl chloride and antimony chloride imparts durable flame resistance to cellulosic fabrics.(83) The formulation can be prepared by reacting titanium tetrachloride and antimony oxide.

$$6\ TiCl_4 + Sb_4O_6 \rightarrow 6\ TiOCl_2 + 4SbCl_3$$

It is necessary that the flame retardant formulation be strongly acidic to prevent hydrolysis of the titanium oxychloride and antimony trichloride. The method of application is by padding fabric through the treating solution and then neutralizing it with a strong solution of sodium carbonate. Titanium and antimony are not merely precipitated within the fibre as oxides; they are thought to form a compound with the cellulose. The fixation of the titanium and antimony is very efficient when about 15 per cent sodium carbonate solution is used to neutralize the solution in the fabric. Cotton fabrics having add-ons of 10 to 15 per cent pass the vertical flame test. Although flame resistance is good, the afterglow lasts several minutes. Breaking strength is usually about 90 per cent or more of the original untreated fabric, and hand is very good. Fabric treated with this process will withstand repeated laundering and drycleaning.

Application of Flame Retardants from Solvent Systems

With pressures being applied from various sources to reduce stream pollution, research is being carried out to find flame retardants that can be applied to textiles from solvent systems. Symm[84] used chlorinated solvents to apply flame retardants including APO and brominated hydroxyl-containing compounds (tribromopentyl alcohols and phosphonates). The compounds in solvent systems were applied to fabric and cured at 160°–170°C to produce flame retardant fabrics with less add-on than is usually required from aqueous systems.

REFERENCES

F. H. Holmes and C. J. G. Shaw, *J. Appl. Chem.* (*London*), **11,** 210–216 (1961).
2. J. L. Gay-Lussac, *Ann. Chim. et Phys.*, **18** (2), 211–216 (1821).
3. M. Leatherman, U.S. Dept. of Agriculture Circular No. 466 (March 1938).
4. H. Schuyten, J. Weaver and J. D. Reid, "Some Theoretical Aspects of the Flameproofing of Cellulose," *Advances in Chemistry, No. 9*, Academic Press, New York, New York, 1954, pp. 7–20.
5. S. L. Madorsky, V. E. Hart and S. Staus, *J. Res. National Bur. Stds.*, **56,** 343 (1956).
6. A. S. Endler and M. L. Hurwitz, *Amer. Dyest. Rep.*, **56** (19), 694–697 (1967).
7. R. B. LeBlanc, *Text. Chem. Color.*, **2** (3), 47–50 (1970).
8. Federal Test Method Std. No. 191, Method 5903, Dec. 1968.
9. "Flame Retardant Properties of Treated Fabric," ASTM D626–55T, American Society for Testing Materials (1965).
10. "Standards for Fireproof Textiles," NFPA 701, PR National Fire Protection Association (1965).
11. "Outline of Proposed Investigations of Factors, Treated Flame-Resistant Clothing, Bedding and Drapery Fabrics," Underwriters Laboratory, Subject 214, Jan. 1964.
12. E. P. Martin, "General Study of the Present Situation in the Field of Flameproof Finishing of Textiles and a Critical Examination of Existing Texting Methods," gdi Occasional Publications No. 45, 41–62 (1969); "Flammability of Fabrics," *Ciba Review*, **4,** 44–46 (1969).
13. P. M. Hay, *Amer. Dyest. Rep.*, **53** (19), 23–26 (1964).
14. Federal Aviation Agency (FAA) Flame Retardant Test.
15. General Motors Corporation Test Method 30–27, 4th Revision, Nov. 13, 1967; Ford Motor Company, Physical Test Method BN24–1, Oct. 30, 1962; Chrysler Corporation Laboratory Procedure LP–463KC–13–01, Nov. 14, 1968.
16. W. A. Reeves, O. J. McMillan, Jr. and J. D. Guthrie, *Text. Res. J.*, **23,** 527–532 (1953).
17. J. V. Beninate, R. M. Perkins, G. L. Drake, Jr. and W. A. Reeves, *Text. Res. J.*, **39** (4), 368–374 (1969).
18. C. P. Fenimore and F. J. Martin, *Mod. Plast.*, **43,** 141, 146, 148, 192 (1966).
19. J. S. Isaacs, *J. Fire and Flammability*, **1,** 36–47 (1970).
20. G. C. Tesoro and C. H. Meiser, Jr., as reported at the TRI Annual Meeting, April 1969.
21. J. J. Willard and R. E. Wondra, *Text. Res. J.*, **40** (3), 203–210 (1970).
22. H. Stepniczka and J. DiPietro, *Smoke Density and LOI-Studies of Polymeric Systems*, in Polymer Conference Series, Univ. of Detroit, 1970, Detroit Col. of Eng. Polymer Inst. 38 pp., processed.
23. B. Miller, *New Approaches to Flame Propagation*, in Polymer Conference Series, Univ. of Detroit, 1970, Detroit Col. of Eng. Polymer Inst. 17 pp., processed.
24. G. Tesoro, *Information Council of Fabric Flammability Proc.*, 173–181 (1968).

25. British Standards 476, Part 2, 1955
26. G. Schon, *Melliand Textilber*, **48,** 215–217 (1967).
27. P. St. Hilaire, K. Knoettner and L. E. Rossiter, *Amer. Dyest. Rep.*, **57,** 942–945 (1968).
28. "Flammability of Clothing Textiles," Text Method 33–1962, *American Association of Textile Chemists and Colorists Technical Manual*, Howes Publishing Co., Inc., New York, N.Y., 1968, pp. B–139–142.
29. Commercial Standards 191–53 (revised) U.S. Dept. of Commerce, Superintendent of Documents, U.S. Gov. Printing Office, Washington, D.C.
30. Flammable Fabrics Act, Public Law 88, 83rd Congress, Chapter 164, HR5069, July 1, 1954, amended July 1, 1958.
31. Federal Specification CCC–T–191b, Federal Supply Service Administration, May 15, 1951, Methods 5908 and 5910.
32. "First Proposed Flammability Standard as Issued by Dept. of Commerce," *Text. Chem. Color.*, **2** (1), 13–16 (1970).
33. *J. Text. Inst.*, **59** (1), 46–57 (1968).
34. J. E. Ramsbottom and A. W. Snoad, "Second Report of the Fabrics Coordinating Research Committee, Dept. of Scientific and Industrial Research," Great Britain (1930).
35. Joseph Bancroft and Sons Co., Brit. Pat. 604197 (1948).
36. F. V. Davis, J. Findley and E. Rogers, *J. Text. Inst.*, **40,** T839–T854 (1949).
37. T. H. Marton (to Courtaulds Ltd.), Brit. Pat. 634690 (1950).
38. American Cyanamid Co., Brit. Pat. 638434 (1950).
39. H. H. Bowen, V. H. Majerus, W. S. Meals and S. Kellett, Brit. Pat. 586095 (1947).
40. S. J. O'Brien, *Text. Res. J.*, **38,** 256–266 (1968).
41. M. L. Nielsen, *Text. Res. J.*, **27,** 603–610 (1957).
42. W. A. Reeves, R. M. Perkins, B. Piccolo and G. L. Drake, Jr., *Text. Res. J.*, **40** (3), 223–231 (1970).
43. J. G. Frick, W. A. Reeves and J. D. Guthrie, *Text. Res. J.*, **27,** 294–299 (1957).
44. J. G. Frick, Jr., J. W. Weaver and J. D. Reid, *Text. Res. J.*, **25,** 100–105 (1955).
45. C. Hamalainen and J. D. Guthrie, *Text. Res. J.*, **26,** 141–144 (1956).
46. L. E. A. Godfrey (assigned to FMC Corp.) U.S. Pat. 3505087 (1970).
47. A. D. F. Toy and Rattenbury, U.S. Pat. 2735789 (1956); J. R. Costello, Jr. and T. P. Traise, U.S. Pat. 2841507 (1958); A. D. F. Toy, R. S. Cooper and T. P. Traise, U.S. Pat. 2867548 (1959).
48. Joseph Bancroft and Sons Co., Brit. Pat. 604197 (1948).
49. F. V. Davis, J. Findley and E. Rogers, *J. Text. Inst.*, **40,** T839–T854 (1949).
50. S. J. O'Brien, *Text. Res. J.*, **38,** 256–266 (1968).
51. M. L. Nielson, *Text. Res. J.*, **27,** 603–610 (1957).
52. L. H. Chance, G. L. Drake, Jr. and W. A. Reeves, U.S. Pat. 3403044 (1968).
53. R. Fikentscher, H. Brandeis and H. Bille, Ger. Pat. 1126259 (1968).
54. L. H. Chance, J. Warren and J. D. Guthrie, U.S. Pat 2743232 (1956).
55. G. L. Drake, Jr., W. A. Reeves and J. D. Guthrie, *Text. Res. J.*, **29,** 270–275 (1959).

56. L. H. Chance, W. A. Reeves and G. L. Drake, Jr., *Text. Res. J.*, **35** (4), 291–298 (1965).
57. R. Aenishänslin, C. Guth, P. Hofmann, A. Maeder and N. Nachbur *Text. Res. J.*, **39** (4), 375–381 (1969).
58. G. C. Tesoro, S. B. Sello and J. J. Willard, *Text. Res. J.*, **38** (3), 245–55 (1968).
59. J. P. Moreau and L. H. Chance, *Amer. Dyest. Rep.*, **59** (5), 37–38, 64 (1970).
60. L. H. Chance and J. P. Moreau, presented at the 159th ACS Meeting, Div. of Cellulose, Wood, and Fiber Chemistry, Houston, Texas, Feb. 22–27, 1970.
61. A. Hoffman, *J. Amer. Chem. Soc.*, **43,** 1684–1688 (1921).
62. F. F. Fynn, W. A. Reeves and J. D. Guthrie, U.S. Pat. 2743299 (1956).
63. J. D. Guthrie, G. L. Drake, Jr. and W. A. Reeves, *Amer. Dyest. Rep.*, **44** (10), 328–332 (1955).
64. *Text. Ind.* (*Atlanta*), **124** (5), 182–185 (1960).
65. B. R. LeBlanc, *Text. Ind.* (*Atlanta*), **132** (10), 274, 296, 298 (1968).
66. G. M. Wagner (to Hooker Chemical Corp.), U.S. Pat. 3243391 (1966).
67. Piedmont Section, AATCC, *Amer. Dyest. Rep.*, **57,** 373–377 (1968).
68. Harold Coates (to Albright and Wilson, Ltd.), U.S. Pat. 3096201 (1963).
69. J. V. Beninate, E. K. Boylston, G. L. Drake, Jr. and W. A. Reeves, *Text. Ind.* (*Atlanta*), **131** (11), 110–114 (1967).
70. J. V. Beninate, E. K. Boylston, G. L. Drake, Jr. and W. A. Reeves, *Amer. Dyest. Rep.*, **57,** 981–985 (1968).
71. J. V. Beninate, E. K. Boylston, G. L. Drake, Jr. and W. A. Reeves, *Text. Res. J.*, **38,** 267–272 (1968).
72. D. J. Donaldson and D. J. Daigle, *Text. Res. J.*, **39** (4), 363–367 (1969).
73. D. J. Daigle and D. J. Donaldson, *Text. Chem. Color.*, **1** (24), 34–36 (1969).
74. F. L. Normand, D. J. Donaldson and G. L. Drake, Jr., *Text. Ind.*, **134** (6), 169–170, 176, 186, 188 (1970).
75. G. L. Drake, Jr., L. H. Chance, J. V. Beninate and J. D. Guthrie, *Amer. Dyest. Rep.*, **51,** 272–274 (1962).
76. W. A. Reeves, G. L. Drake, Jr., L. H. Chance and J. D. Guthrie, *Text. Res. J.*, **27,** 260–266 (1957).
77. T. D. Miles, F. A. Hoffman and A. Merola, *Amer. Dyest. Rep.*, **49,** 596–599 (1960).
78. R. B. LeBlanc, *Text. Res. J.*, **35,** 341–346 (1965).
79. R. J. Brysson, B. Piccolo and A. M. Walker, "Calcium-Phosphorus Deposition During Home Laundering," (in manuscript).
80. W. A. Reeves, "Some Factors Influencing the Effectiveness of Durable Phosphorus-Containing Flame Retardants for Cellulosic Textiles," presented at the 160th National ACS Meeting, Division of Cellulose, Wood, and Fiber Chemistry, Chicago, Ill., Sept. 13–18, 1970.
81. E. C. Clayton and L. L. Heffner (to William E. Hooper and Sons, Inc.), U.S. Pat. 2299612 (1942).
82. R. B. LeBlanc, *Text. Ind.* (*Atlanta*), **132** (10), 274, 294, 296, 298 (1968).
83. H. C. Gulledge and G. H. Seidel, *Ind. and Eng. Chem.*, **42,** 440 (1950).
84. R. H. Symm, *Text. Chem. Color.*, **1** (6), 42–45 (1969).

INDEX